Chapter 1: Understanding Water Scarcity and Waste

Water scarcity is a growing concern worldwide, affecting millions of people and ecosystems. The problem is exacerbated by the inefficient use and management of water resources. Scarcity results from a combination of natural factors and human activities, including climate change, population growth, and poor water management practices. As the demand for water increases, the availability of clean, fresh water decreases, leading to significant environmental and social impacts.

Water waste occurs when water is used inefficiently or unnecessarily, leading to a depletion of available resources. This waste can happen at various levels, including industrial processes, agricultural practices, and domestic usage. Understanding the causes and consequences of water waste is crucial for developing effective strategies to address the issue. By examining the factors contributing to water scarcity and waste, we can identify areas for improvement and implement solutions to conserve and manage water more effectively.

Climate change plays a significant role in water scarcity, affecting precipitation patterns and water availability. As temperatures rise and weather patterns become more unpredictable, regions may experience more frequent droughts or heavy rainfall, impacting water supply and demand. Additionally, population growth and urbanization put additional pressure on water resources, as more people require access to clean water for drinking, sanitation, and other needs.

Inefficient water use in agriculture is a major contributor to water waste. Traditional irrigation methods, such as flood irrigation, can result in significant water losses due to evaporation and runoff. Modern techniques, such as drip irrigation and precision farming, offer more efficient ways to deliver water directly to crops, reducing waste and improving water use efficiency.

In industrial settings, water is often used in large quantities for manufacturing processes, cooling systems, and waste disposal. Many industries have implemented water recycling and reuse practices to minimize waste and reduce their environmental impact. However, there is still considerable room for improvement in many sectors to achieve more sustainable water use.

Domestic water usage also contributes to water waste, with common practices such as leaving taps running, overwatering lawns, and inefficient appliances leading to unnecessary water consumption. Educating the public about water conservation and promoting the adoption of water-saving technologies can help reduce domestic water waste and contribute to overall water efficiency.

Addressing the problem of water scarcity and waste requires a comprehensive approach that includes understanding the underlying causes, implementing effective management strategies, and promoting responsible water use at all levels. By taking action to reduce water waste and improve water management, we can work towards a more sustainable and equitable future for all.

Chapter 2: Industrial Water Use and Inefficiency

Industrial water use is a significant factor in the global water scarcity crisis. Industries consume vast amounts of water for various processes, including manufacturing, cooling, and waste disposal. Inefficient water use and poor management practices in the industrial sector can contribute to water wastage and environmental degradation.

Industries often rely on outdated technologies and practices that result in substantial water losses. For example, cooling systems in power plants and manufacturing facilities frequently use large volumes of water for cooling purposes. Many of these systems are designed with limited consideration for water efficiency, leading to significant water losses through evaporation and discharge.

To address industrial water inefficiency, companies can implement several strategies to reduce water consumption and improve water management. One approach is to adopt water recycling and reuse technologies, which involve capturing and treating wastewater for reuse in industrial processes. By recycling water, industries can reduce their reliance on freshwater sources and minimize waste.

Another strategy is to optimize water use through process improvements and technological upgrades. For example, industries can implement water-efficient technologies, such as advanced cooling systems and water-saving equipment, to reduce water consumption. Conducting regular water audits and assessing water use patterns can also help identify areas for improvement and implement targeted solutions.

In addition to technological improvements, industries can adopt best practices for water management and conservation. This includes setting water use reduction targets, monitoring water usage, and engaging in water stewardship initiatives. Collaborating with stakeholders, including governments, NGOs, and local communities, can also help address water-related challenges and promote sustainable water use.

Regulatory frameworks and policies play a crucial role in driving industrial water efficiency. Governments can implement regulations and incentives that encourage industries to adopt water-saving practices and technologies. For example, policies that promote water efficiency standards and provide financial incentives for water conservation projects can motivate industries to invest in sustainable water management practices.

Despite the potential benefits of improved water management in industries, challenges remain. In many regions, industries face barriers such as limited access to technology, high costs of implementation, and lack of awareness about water efficiency. Addressing these challenges requires coordinated efforts between governments, industries, and other stakeholders to support the adoption of water-saving practices and technologies.

Despite the potential benefits of improved water management in industries, challenges remain. In many regions, industries face barriers such as limited access to technology, high costs of implementation, and lack of awareness about water efficiency. Addressing these challenges requires coordinated efforts between governments, industries, and other stakeholders to support the adoption of water-saving practices and technologies.

Chapter 3: Agricultural Water Management Challenges

Agriculture is a major consumer of water, with irrigation accounting for a significant portion of global freshwater use. However, traditional irrigation practices often lead to substantial water wastage and inefficiencies. Addressing these challenges requires adopting modern water management techniques and improving overall agricultural practices.

Traditional irrigation methods, such as flood irrigation, involve applying large volumes of water to fields, often resulting in significant losses due to evaporation and runoff. These practices can lead to overuse of water resources and contribute to water scarcity. Modern irrigation techniques, such as drip irrigation and sprinkler systems, offer more efficient alternatives by delivering water directly to plant roots and reducing water wastage.

Drip irrigation systems use a network of tubes and emitters to provide a slow, steady supply of water to plants. This method minimizes water loss through evaporation and runoff and allows for precise control over water application. Drip irrigation is particularly beneficial for crops in arid and semi-arid regions, where water resources are limited.

Sprinkler systems, on the other hand, use a network of pipes and nozzles to distribute water over crops in a controlled manner. Modern sprinkler systems can be designed to minimize water wastage by adjusting the spray pattern and applying water only when needed. Implementing efficient sprinkler designs and scheduling irrigation based on crop needs can help reduce water use and improve overall efficiency.

In addition to modern irrigation techniques, implementing soil moisture management practices can further enhance water use efficiency in agriculture. Techniques such as mulching, which involves covering soil with organic or inorganic materials, can help retain soil moisture and reduce the need for frequent irrigation. Soil moisture sensors can also provide real-time data on soil conditions, allowing farmers to optimize irrigation schedules and avoid overwatering.

Water conservation practices in agriculture also involve improving water storage and management infrastructure. Building reservoirs, ponds, and rainwater harvesting systems can help capture and store water for use during dry periods. Efficient water storage and distribution systems can reduce water losses and ensure a more reliable water supply for agricultural activities.

Education and training for farmers are essential for promoting the adoption of water-efficient practices. Providing access to information on modern irrigation techniques, soil moisture management, and water conservation strategies can help farmers make informed decisions and improve their water use practices. Extension services and support programs can play a crucial role in disseminating knowledge and promoting best practices in agricultural water management.

Addressing agricultural water management challenges requires a comprehensive approach that includes adopting modern irrigation techniques, improving soil moisture management, and investing in water storage and distribution infrastructure. By implementing these strategies, farmers can reduce water wastage, enhance water use efficiency, and contribute to more sustainable agricultural practices.

Chapter 4: Domestic Water Waste and Conservation

Domestic water use accounts for a significant portion of global water consumption, and inefficient practices can lead to substantial water wastage. Addressing domestic water waste requires a combination of behavioral changes, technology adoption, and public awareness to promote water conservation and ensure responsible water use.

One common source of domestic water waste is the use of outdated plumbing fixtures and appliances. Traditional toilets, faucets, and showerheads often use more water than necessary, contributing to excessive water consumption. Upgrading to water-efficient fixtures and appliances, such as low-flow toilets, aerated faucets, and water-saving showerheads, can significantly reduce water use without compromising performance.

Water leaks are another major issue contributing to domestic water waste. Leaky faucets, toilets, and pipes can result in significant water loss over time. Regular maintenance and prompt repairs of plumbing fixtures are essential for preventing and addressing leaks. Simple measures, such as checking for leaks and fixing them promptly, can help conserve water and reduce utility bills.

Water usage habits also play a crucial role in domestic water conservation. Common practices, such as leaving the tap running while brushing teeth, overwatering lawns, and taking long showers, can lead to unnecessary water consumption. Adopting water-saving habits, such as turning off the tap while brushing, using a broom instead of a hose for cleaning driveways, and taking shorter showers, can contribute to significant water savings.

Efficient landscaping practices can also help reduce water use in domestic settings. Implementing drought-tolerant plants, using mulch to retain soil moisture, and employing efficient irrigation techniques can minimize water consumption for outdoor areas. Choosing native plants that require less water and adapting landscaping practices to local climate conditions can further enhance water conservation efforts.

Public awareness and education are critical for promoting domestic water conservation. Informing individuals about the importance of water conservation, providing tips for reducing water use, and encouraging the adoption of water-saving technologies can foster a culture of responsible water use. Community initiatives, such as water conservation campaigns and educational programs, can help raise awareness and drive positive change.

Government policies and regulations can also support domestic water conservation efforts. Implementing water use restrictions during drought periods, providing incentives for water-efficient home improvements, and promoting water-saving technologies can encourage households to adopt conservation practices. Collaboration between government agencies, utilities, and communities can help develop and implement effective water conservation programs.

By addressing domestic water waste through a combination of technology adoption, behavioral changes, and public awareness, individuals can contribute to more sustainable water use and help address the global water scarcity crisis. Implementing water-saving practices and promoting responsible water use at the household level can have a significant impact on overall water conservation efforts.

Chapter 5: The Role of Technology in Reducing Water Waste

Technology plays a vital role in addressing water waste and improving water management practices. Advances in technology have provided innovative solutions for monitoring, conserving, and managing water resources more effectively. From smart irrigation systems to water-efficient appliances, technology offers numerous opportunities to reduce water wastage and enhance overall water use efficiency.

Smart irrigation technology has revolutionized water management in agriculture and landscaping. These systems use sensors, weather data, and advanced algorithms to optimize irrigation schedules and deliver water only when needed. By monitoring soil moisture levels and adjusting irrigation based on real-time data, smart irrigation systems can significantly reduce water use and minimize waste.

Water-saving appliances, such as low-flow toilets, faucets, and showerheads, have also made a significant impact on domestic water conservation. These appliances are designed to use less water while maintaining functionality and performance. For example, low-flow toilets use less water per flush compared to traditional models, and water-efficient showerheads reduce water flow without compromising the shower experience.

Leak detection technology is another area where advancements have helped address water waste. Smart water meters and leak detection systems can monitor water usage in real-time, identify leaks, and alert homeowners or facility managers to potential issues. Early detection of leaks can prevent significant water loss and reduce the need for costly repairs.

In industrial settings, water recycling and reuse technologies have become increasingly important. Industries are implementing systems to capture and treat wastewater for reuse in various processes, reducing their reliance on freshwater sources. Technologies such as reverse osmosis, membrane filtration, and ultraviolet disinfection are used to purify and recycle water, minimizing waste and environmental impact.

Water management software and data analytics tools are also playing a crucial role in optimizing water use. These tools can analyze water usage patterns, identify inefficiencies, and provide recommendations for improvement. By leveraging data and technology, organizations can make informed decisions about water management and implement strategies to reduce waste.

Research and development in water technology continue to drive innovation and improve water conservation efforts. Emerging technologies, such as nanotechnology and advanced materials, offer new possibilities for water treatment and efficiency. For example, nanofiltration membranes can remove contaminants at the molecular level, providing cleaner water with reduced waste.

The integration of technology into water management practices requires investment and commitment from various stakeholders, including governments, industries, and individuals. Supporting research, investing in infrastructure, and promoting the adoption of water-saving technologies can help accelerate progress and address the global water scarcity crisis.

In summary, technology plays a crucial role in reducing water waste and improving water management. By adopting innovative solutions and leveraging advancements in technology, we can enhance water efficiency, conserve resources, and contribute to a more sustainable future.

Chapter 6: Water Conservation Policies and Regulations

Water conservation policies and regulations are essential for addressing water scarcity and promoting sustainable water use. Governments and organizations implement various policies and regulations to manage water resources effectively, reduce waste, and ensure equitable access to clean water. Understanding and supporting these policies can help drive positive change and improve overall water management.

One key area of water conservation policy is regulating water usage in different sectors, including agriculture, industry, and domestic settings. Policies may include restrictions on water use during drought periods, incentives for water-efficient technologies, and requirements for water conservation measures. For example, some regions have implemented water use restrictions that limit outdoor watering during dry conditions to conserve water resources.

Another important aspect of water conservation policy is promoting water-efficient technologies and practices. Governments may offer financial incentives, such as rebates or tax credits, to encourage the adoption of water-saving appliances and systems. Policies that require the installation of low-flow fixtures, efficient irrigation systems, and water-saving technologies can help reduce overall water consumption.

Water pricing and cost recovery mechanisms are also integral to water conservation policies. Implementing tiered pricing structures that charge higher rates for increased water use can incentivize users to conserve water and reduce waste. Cost recovery mechanisms, such as charging for water services based on actual usage, can help fund water infrastructure projects and ensure sustainable management of water resources.

Public education and awareness campaigns are crucial for promoting water conservation and encouraging responsible water use. Policies that support community outreach, educational programs, and public awareness initiatives can help inform individuals about the importance of water conservation and provide practical tips for reducing water use. Engaging the public in water conservation efforts fosters a culture of responsibility and helps drive collective action.

Collaborative efforts between governments, organizations, and communities are essential for effective water conservation policies. Stakeholders can work together to develop and implement policies that address local water challenges, share best practices, and support innovative solutions. Collaboration can enhance policy effectiveness and ensure that conservation efforts are aligned with regional needs and priorities.

Monitoring and enforcement are critical components of water conservation policies. Governments and regulatory agencies need to monitor water use, assess the effectiveness of policies, and enforce compliance with regulations. Regular reporting, inspections, and penalties for non-compliance can help ensure that policies are implemented effectively and that water resources are managed sustainably.

In summary, water conservation policies and regulations play a vital role in addressing water scarcity and promoting sustainable water use. By supporting effective policies, adopting water-efficient technologies, and engaging in public education, we can work towards a more sustainable and equitable future for our water resources.

Chapter 7: Community Initiatives and Water Conservation

Community initiatives play a crucial role in promoting water conservation and addressing local water challenges. By engaging individuals, organizations, and local governments, communities can implement effective water-saving strategies, raise awareness, and drive positive change. Community-based efforts can complement broader policy measures and contribute to more sustainable water management practices.

One example of a successful community initiative is the establishment of water conservation programs and campaigns. These programs often involve educational workshops, outreach events, and public awareness campaigns to inform residents about the importance of water conservation and provide practical tips for reducing water use. Community organizations, schools, and local governments can collaborate to deliver these programs and reach a wide audience.

Community-based water conservation projects can also focus on specific local issues, such as improving water infrastructure or addressing water pollution. For example, local groups may organize clean-up events to remove debris from water bodies, restore wetlands, or advocate for better wastewater treatment facilities. These projects not only improve local water quality but also foster community engagement and collaboration.

Another approach to community water conservation is the implementation of water-saving technologies and practices at the local level. Communities can work together to install water-efficient fixtures, such as low-flow toilets and faucets, in public buildings and shared spaces. Additionally, community gardens and green spaces can adopt sustainable irrigation practices and drought-tolerant landscaping to reduce water consumption.

Community engagement and participation are essential for the success of water conservation initiatives. Involving residents in decision-making processes, soliciting feedback, and encouraging active participation can help ensure that conservation efforts address local needs and priorities. Engaging community members as volunteers, advocates, and partners can enhance the effectiveness of water conservation projects and foster a sense of shared responsibility.

Local governments and organizations can also support community initiatives by providing resources, funding, and technical assistance. Grants, subsidies, and other forms of support can help communities implement water-saving technologies, conduct educational programs, and address water-related challenges. Collaboration between government agencies, non-profits, and community groups can amplify the impact of conservation efforts and drive meaningful change.

Monitoring and evaluating the effectiveness of community initiatives are important for assessing progress and identifying areas for improvement. Collecting data on water use, conservation outcomes, and participant feedback can help measure the success of projects and inform future efforts. Regular reporting and communication with community members can also maintain engagement and support for ongoing conservation activities.

In summary, community initiatives play a vital role in promoting water conservation and addressing local water challenges. By engaging residents, implementing water-saving technologies, and supporting collaborative projects, communities can contribute to more sustainable water management and drive positive change. Community-based efforts complement broader policy measures and foster a culture of responsible water use.

Chapter 8: Global Water Scarcity and Its Impact

Global water scarcity is a critical issue affecting billions of people and ecosystems worldwide. The problem of water scarcity arises from a combination of natural and human factors, including climate change, population growth, and inefficient water use. Understanding the impact of water scarcity on different regions and communities is essential for developing effective solutions and promoting sustainable water management.

The impact of global water scarcity is far-reaching, affecting various aspects of human life and the environment. In many regions, water scarcity leads to inadequate access to clean drinking water, resulting in health problems, including waterborne diseases and malnutrition. Lack of access to safe water also affects sanitation and hygiene practices, further exacerbating health risks.

Agriculture is another sector significantly impacted by water scarcity. As a major consumer of freshwater resources, agriculture relies on a reliable water supply for crop irrigation and livestock management. Water scarcity can lead to reduced agricultural productivity, food insecurity, and economic losses for farmers. In regions facing severe water shortages, farmers may struggle to grow crops and maintain livestock, impacting local food supplies and livelihoods.

Water scarcity also affects economic development and infrastructure. In areas with limited water resources, industries may face challenges in sourcing water for manufacturing processes, cooling systems, and waste disposal. The resulting constraints on industrial activities can hinder economic growth and reduce opportunities for employment and investment.

Ecosystems and natural habitats are also affected by global water scarcity. Rivers, lakes, and wetlands rely on consistent water flows to maintain ecological balance and support diverse plant and animal species. Water scarcity can lead to the degradation of these habitats, loss of biodiversity, and disruptions to ecosystem services, such as water filtration and flood regulation.

Addressing global water scarcity requires a multi-faceted approach that includes improving water management practices, investing in infrastructure, and promoting sustainable water use. Solutions may involve implementing water-saving technologies, enhancing water recycling and reuse, and adopting efficient irrigation methods. Additionally, raising awareness about water conservation and advocating for policy changes can contribute to more effective water management and reduced scarcity.

International cooperation and collaboration are essential for addressing global water scarcity. Countries and organizations can work together to share knowledge, technologies, and resources to tackle water-related challenges. By fostering global partnerships and supporting initiatives that promote sustainable water use, we can work towards a more equitable and resilient future for our planet's water resources.

In summary, global water scarcity has wide-ranging impacts on human health, agriculture, economic development, and ecosystems. Understanding these impacts and implementing effective solutions are crucial for addressing the water crisis and promoting sustainable water management. By taking action at local, national, and international levels, we can work towards a more sustainable and equitable future for all.

Chapter 9: Innovative Solutions for Water Waste Reduction

Innovative solutions play a key role in addressing the issue of water waste and improving water management. Advances in technology, new approaches to water conservation, and creative problem-solving can help reduce water wastage and promote more efficient use of water resources. Exploring these solutions can provide valuable insights into how we can tackle the global water crisis and achieve sustainable water management.

One of the most promising innovations in water waste reduction is the development of smart water management systems. These systems use sensors, data analytics, and automated controls to monitor and optimize water use in real-time. Smart meters, for example, provide detailed information about water consumption, enabling users to track usage patterns and identify areas for improvement. Automated irrigation systems can adjust watering schedules based on weather conditions and soil moisture levels, reducing water waste in agriculture and landscaping.

Water recycling and reuse technologies are also at the forefront of innovative solutions for reducing water waste. Advanced treatment processes, such as membrane filtration and reverse osmosis, allow for the purification and reuse of wastewater. These technologies can be applied in various settings, including industrial processes, municipal wastewater treatment plants, and residential applications. By recycling and reusing water, we can reduce the demand for freshwater resources and minimize environmental impact.

Green infrastructure and sustainable design practices offer additional solutions for water conservation and waste reduction. Green roofs, rain gardens, and permeable pavements can help manage stormwater runoff and reduce the burden on conventional drainage systems. These features not only improve water management but also provide environmental and aesthetic benefits, such as reducing urban heat islands and enhancing biodiversity.

Behavioral changes and public awareness are also important components of innovative solutions for water waste reduction. Educational campaigns and outreach programs can inform individuals and communities about the importance of water conservation and provide practical tips for reducing water use. Encouraging the adoption of water-saving habits, such as turning off the tap while brushing teeth and fixing leaks promptly, can lead to significant reductions in water wastage.

Collaboration between governments, businesses, and research institutions is essential for fostering innovation in water management. Public-private partnerships can support the development and implementation of new technologies and solutions, while research institutions can contribute to advancements in water science and engineering. By working together, stakeholders can address water-related challenges and drive progress towards more sustainable water use.

In summary, innovative solutions play a critical role in reducing water waste and improving water management. By leveraging technology, adopting green infrastructure practices, promoting behavioral changes, and fostering collaboration, we can make significant strides towards more efficient and sustainable water use. Addressing the issue of water waste requires a comprehensive approach that incorporates creative problem-solving and advances in technology to achieve lasting impact.

Chapter 10: The Impact of Climate Change on Water Resources

Climate change has a profound impact on water resources, affecting the availability, quality, and distribution of water across the globe. Changes in temperature, precipitation patterns, and extreme weather events are altering the way water is managed and utilized, with significant implications for ecosystems, agriculture, and human communities.

One of the primary effects of climate change on water resources is the alteration of precipitation patterns. Increased temperatures can lead to more intense and frequent rainfall in some regions, while others may experience prolonged droughts and reduced precipitation. These changes can affect the availability of freshwater resources, leading to challenges in managing water supply and demand.

Melting glaciers and ice caps are another consequence of climate change that impacts water resources. As temperatures rise, glaciers and ice caps are melting at an accelerated rate, contributing to rising sea levels and altering the flow of freshwater into rivers and lakes. This can lead to changes in water availability, especially in regions that rely on glacier-fed rivers for their water supply.

Extreme weather events, such as hurricanes, floods, and heatwaves, are becoming more common due to climate change. These events can cause significant damage to water infrastructure, contaminate water sources, and disrupt water supply and distribution systems. The increased frequency and severity of extreme weather events pose challenges for managing water resources and ensuring reliable access to clean water.

Climate change also affects the quality of water resources. Rising temperatures can lead to increased evaporation rates, which can concentrate pollutants and contaminants in water sources. Additionally, changes in precipitation patterns can impact the flow and dilution of pollutants, affecting water quality and increasing the risk of waterborne diseases.

Adaptation and mitigation strategies are essential for addressing the impact of climate change on water resources. Implementing water conservation measures, improving water infrastructure, and adopting sustainable water management practices can help reduce vulnerability to climate change and enhance resilience. Additionally, investing in research and monitoring systems can provide valuable data for understanding and addressing the effects of climate change on water resources.

International cooperation and collaboration are crucial for addressing the global impacts of climate change on water resources. Countries can work together to share knowledge, technologies, and best practices for managing water in a changing climate. By fostering global partnerships and supporting climate adaptation initiatives, we can work towards a more resilient and sustainable future for our water resources.

In summary, climate change has significant implications for water resources, affecting precipitation patterns, water quality, and infrastructure. Understanding and addressing these impacts through adaptation and mitigation strategies is essential for managing water resources in a changing climate and ensuring sustainable access to clean water.